Time Restricted Eating

A Look Into the Lifestyle

Dr. Julie Wei-Shatzel

LifeRich Publishing is a registered trademark of The Reader's Digest Association, Inc.

LifeRich Publishing books may be ordered through booksellers or by contacting:

LifeRich Publishing
1663 Liberty Drive
Bloomington, IN 47403
www.liferichpublishing.com
844-686-9607

Cover designed by Blair Shatzel

ISBN: 978-1-4897-3662-8 (sc)
ISBN: 978-1-4897-3663-5 (hc)
ISBN: 978-1-4897-3664-2 (e)

Library of Congress Control Number: 2021912607

Print information available on the last page.

LifeRich Publishing rev. date: 06/24/2021

Contents

Introduction

What Medicine is Missing

I walked down the empty hospital corridor, the bustle and activity of the day having vanished as night set in. All of the hospital entrance doors were locked. At this hour, the only way for non-medical staff to enter was through the Emergency Room. As a medical resident, my duties at the hospital ran in shifts covering day and night; sometimes three days in a row nonstop. Medical residents in teaching hospitals across the country often have to manage crisis emergency events for patients both during the day as well as through the long hours of the night.

On this particular night, I was the intern responsible for all so-called "no-doc" deliveries by pregnant mothers and admissions of sick patients—those without assigned physicians. I would essentially be in charge of taking care of these patients from the moment of their admission throughout the duration of their hospital stay. The length of these night shifts—and of all of the shifts given to residents—made certain that we would get exposure to a wide range of cases at the hospital, from the relatively uncomplicated to the most complicated conditions.

I unlocked the door to an inconspicuous room—it was unoccupied. The hospital had allocated one of its patient rooms to the residents, so that we would have a place to access a computer, catch a few winks, or even eat, if time allowed. The room was a duplicate of the patient rooms; the bed comprising of a floppy mattress with metal coils, sealed in green vinyl plastic, and covered in thin sheets. This hospital-issued bed bore no resemblance to my bed at home. An unused nightstand and a rolling meal tray filled the rest of the space. If I ever did manage to catch a few minutes of sleep in the late hours of the night, the sounds of telemetry monitors, various beeps, and the clamoring of clogs outside the door always made sure to keep me at stage one, non-restorative sleep.

This shift would last 36 hours or more, depending on the number of admissions through the ER during the evening and early morning. The more admissions, the larger the load of ill patients I would oversee. They would need tests ordered, ranging from imaging to lab work, and then be provided with medication and fluids. Residents in my training program knew both the night crew and the day crew at the hospital. We worked with the day and night shift nursing staff, we knew the janitor, and we knew all the kitchen staff. The night shift was like its own small community.

Covering the evening shift at the hospital was as busy as during the day, but in a different way. There were fewer other teams of residents there to provide support, leaving the night resident to cover all the patients of the other residents. Then, the shift itself was not completely over until we were relieved by another resident coming on shift, giving each of us about a day to recover before we went right back into another shift at the hospital. Fatigue was an obstacle, but vending machine coffee was always available.

Our night shifts ran into the mornings. A patient's morning vitals were documented by the nursing staff by 5 am, while imaging reports were dictated by the radiologists all through the night. Every morning, blood test results were available by 6:30 am, the phlebotomy staff having drawn the blood just two hours before. We tracked everything about each patient and paid special attention to their progress, with the ultimate goal of their recuperation so they could return home.

On a typical night, checking on our resident team's patients meant walking up and down numerous flights of stairs and through extensive corridors to reach each of them in their rooms dispersed throughout the hospital. If a code was called over the hospital paging system, I had to drop what I was doing, rush to the patient room, and run the code with the team. There was always a slight feeling of trespass, of being allowed into places that were off-limits to the outside world but to which those in our small community were privy. My life then, as with every other resident in my training program, was centered around caring for patients in the hospital. We cheered at their improvements and solemnly monitored their declines.

Four years of residency training took me to hospitals in Michigan and California, and allowed me the opportunity to work with amazing physicians in the field. I worked with specialists who had trained at premier institutions, and who had themselves trained with premier mentors in their specialized fields. By the time I graduated, I had accumulated 8 years of medical training, including medical school, and as a typical graduate going into my first job in medicine, I believed I knew a lot.

Clinical practice—where patients are seen in the office for scheduled visits—was a stark contrast to the hospital training I had experienced during the last three years of my residency. I faced new and different challenges. Unlike my previous hospitalized patients, my new patients did not have acute medical issues. Rather, they often had chronic medical issues related to cholesterol, blood pressure, and blood sugar. These were problems that could be managed, improved, and at best reversed to prevent eventual hospitalization. The completion of my training was therefore marked by a transition of my responsibilities from hospital care, to caring for patients in the community clinic; from treatment of the very ill, to prevention of illness. What I soon found as a family practice physician was that the major challenges facing the relatively healthy patients in my community stemmed from one disorder: that of being mildly or moderately overweight.

In all those years of my medical training, after all that time working with highly experienced and skilled physicians, I learned that nutrition, diet, and weight loss were not on the spectrum of medical treatments that we doctors are trained to support. I still recall the day when I asked my attending physician how to guide patients who ask how to lose weight. "Weight Watchers," he had answered. "They have the longest track record." And that was the end of the discussion. It was the best answer he had for a problem that he knew even "Weight Watchers" could not permanently fix.

The whole topic of weight loss and battling obesity seemed to be an issue that my mentors found uninteresting. It simply did not generate the same energy and interest as more complex kinds of medical decision making. So, from that time on, my impression was that weight issues had to do with calories—because isn't that what weight loss programs taught? Caloric reduction, along with coaches advising members to stick to caloric reduction, along with peers who also focused on their personal struggles and triumphs with caloric reduction.

I even observed some physicians merely provide diet and nutritional handouts to their patients in place of counseling. I recall seeing one patient receive such an informational handout from a doctor, and the disappointed look on her face as she read the content describing a nutritional food pyramid. Silently, the medical community was placing the blame on their patients; for their lack of discipline in making the right "lifestyle changes."

In the 1990s, there was a new diet that moved beyond the concept of just reducing calories for weight loss. This was the Atkins diet: repurposed from decades before and popularized by its extreme elimination of carbohydrate calories. It utilizes a starvation protection pathway, one which allows the body to burn fat for fuel instead of the usual and preferred fuel of glucose. This protection pathway only kicks in when there is no glucose around to be used, and acts as a sort of survival route for the body to allow it to still run its essential organ processes and avoid a complete shut down. Kicking in ketosis, this pathway is an evolutionarily protective one, possibly developed as an alternative regulator of metabolism during times of famine. By eating mostly fat and protein, followers of this diet restrict carbohydrate intake in order to push their body into a state of ketosis, or a state where fat is burned for fuel.

This change in diet composition may sound familiar. Today, it is similar in strategy to the "Keto" diet. The approach was intriguing to me, since this particular change added additional complexity to our understanding of metabolism. This pathway implied that the "calories in, calories out" formula was overly simplistic. Many people practiced Atkins in the 1990s—indeed, even some of my physician colleagues had used this program to lose weight. I observed some dramatic weight loss successes, but also that few could stay on the program for more than a year. Most participants seemed to gradually gain back their weight over time, and it did not appear to be a long-term lifestyle change that was sustainable over a lifetime. Neither caloric reduction diets nor composition change diets could be used to easily overcome obesity.

I was curious to know if we could do better. Was there another more manageable approach out there? I would not come across the answer until the field of circadian biology made its dramatic breakthroughs over the last decade.

The basis of most medical school training is the information found in textbooks, with little emphasis on current academic research happening in real time. The two fields are widely separated: medicine on one side and academic scientific research on the other. Research makes breakthroughs every single day, but they may take years—if not decades—to manifest as new treatment protocols that directly benefit patients. Yet, trail-blazing publications coming from the world of academic research are what eventually become known and accepted as common practice. Circadian biology is a great example: it is a relatively young field, but its breakthroughs are already helping us to unlock the complex formula of metabolism.

According to traditional medical textbooks, the basic formula of metabolism is that calories in equals calories out, with a side note mentioning the ketosis pathway. Overall, it implies that obesity and related diseases are caused simply by over-consumption and under-utilization of energy—in other words, it is primarily caused by overeating and a lack of exercise. This formula is simple and exact; there was no way around it. It has been reinforced in the past by popular diets, with their focus on eliminating certain food groups or severely reducing total caloric intake. Indeed, the world of weight loss is and always has been driven by these two long standing beliefs, and current medical advice is largely consistent with them.

I shared this misguided belief right up until the day I came across a study published in 2012 out of Dr. Satchin Panda's lab at the Salk Institute in San Diego. When I read this elegantly designed study, my entire understanding of metabolism was fundamentally changed—and I nearly fell out of my chair when I realized I would never have to forgo cheesecake again!

I hope that the following case studies, as compiled from patients in our collaborative study between Mercy Medical Group and the Salk Institute, will help you to see the outcomes that are possible when this new and groundbreaking research from Dr. Satchin Panda's academic world is applied to real people. What was once deemed impossible is now possible. The simplistic formula previously framing the view of the medical world regarding obesity has become archaic. Patients from our ongoing study illustrate how, when given the right strategy, they can overcome their "diagnoses" and turn them into opportunities. For many, these are significant opportunities to reverse disease, drive up metabolism, and dramatically improve their quality of life.

These few case studies represent what tremendous new possibilities await us, and how a dramatic paradigm shift is in progress when it comes to how obesity and its related diseases can be treated. It is through the application of academic research from teams of circadian biologists like those in Dr. Panda's lab that we have now begun to see what we as physicians have been missing all along: a beautiful internal pathway of timing within each of us that, once harnessed, allows the body itself to do nearly all of the work of enhancing health—with the side benefit of weight loss! The ultimate benefit of increased metabolism, and the improvement of all processes and pathways within the body: this is the potential of Time Restricted Eating.

I hope that these patient cases will inspire you to harness this new knowledge of the circadian system within you, to improve your health, your immune system, and to ultimately reduce unnecessary energy stores that are inhibiting your organs from functioning optimally.

Chapter 1

A Groundbreaking Animal Study

The physician audience drew their attention to the first slide on the screen, depicting an image of two mice side by side. One was obese and the other was thin—the contrast between the two was striking.

"These two mice are actually twins," I said, pointing to the image. "They started life at exactly the same weight and size. In fact, they are actually genetically identical mice."

I was the first speaker of the morning. It was 7:00 am—relatively early for a medical education conference—and these doctors had already had to leave their hotel, drive to the convention center, park their cars, and then walk to the lecture hall to arrive on time. It was also a weekend, and many of them had traveled here with their families. Yet this group of physicians, motivated by their interest in this new and burgeoning field of metabolism chose to attend this lecture on the topic of Time Restricted Eating. They understood that this information could influence the way they practiced medicine. In evaluating my slide, of these two drastically different looking mice, they likely considered the possible differing environmental pressures that these two identical mice may have faced in order to account for such a divergence in size. They may have assumed that the two must have eaten different diets or experienced different levels of activity, for example.

"Over the four months of this animal study, these two mice were fed the exact same diet, and they ate the same amount of food," I explained. "Additionally, not only did these two mice have the same kind of food and the same amount of food, but they were also both fed a terrible diet. In fact, they were both fed the worst kind of food you can imagine, their mouse chow being so high in fat and sugar that it was the equivalent of a human eating potato chips and ice cream for every meal."[1] A silence fell across the audience, and I could see some raised eyebrows from the physicians in the front row.

This image of an obese mouse and a lean mouse demonstrated stark differences in research outcomes for these twins, despite the mice eating the same types and amounts of food—but how could this be possible? This study countered the most widely accepted theory of metabolism—that diet mattered the most. It countered the formula you and I have come to accept as irrefutable, that calories in equals calories out. How was it possible that a mouse could overcome the

---

[1]   Satchidananda Panda, "Time restricted feeding without reducing caloric intake prevents metabolic diseases in mice fed a high fat diet", *Cell Metab. 2012* Jun 6; 15(6): 848–860.

effects of a fast food diet? How did that one mouse manage to remain lean while its counterpart grew obese? The two had eaten the same amount of food for four months, and eaten the same fast food diet for every meal. Logically, one mouse had become obese and suffered the diseases associated with obesity. But illogically, the other mouse had skirted any negative side effects of this diet and suffered no health consequences at all. What had the researchers done?

My physician audience may have assumed that these two twin mice had been fed different amounts of food, or perhaps that one had exercised and the other had not. After all, what do doctors recommend when you ask them about weight loss? Most likely their answer would be "diet and exercise." Physicians typically make the assumption that the barrier to a patient's weight loss is either a lack of activity or poor food choices. I was hoping to dispel this very myth. There is more to metabolism than just diet and exercise, and this mouse study said it all.[2]

This was one of several talks that I had given to medical audiences as a way of sharing a new and groundbreaking finding in the field of human metabolism—a discovery made by academic researchers that revealed the presence of an internal timing system within your cells. This timing system sets the circadian rhythms (or rhythms of activity) within your body, and it can unleash a hidden metabolic pathway. How to harness this pathway involves careful timing

---

[2]   Satchidananda Panda, "Time-Restricted Feeding Prevents Obesity and Metabolic Syndrome in Mice Lacking a Circadian Clock" *Cell Metab.* 2019 Feb 5; 29(2): 303–319.e4.

of two critical activities: eating and fasting. By presenting at medical education conferences and participating in the human aspect of applying this new principle, I hoped to help clear up the medical misunderstanding that obesity is solely a calories and exercise problem. Instead, it is much more likely a problem with the body's internal timing system, which is set by *when* a person eats and fasts.

Has your attention always been fixed on examining your calorie intake and exercise schedule? Over 11,000 research publications conclude that a high fat and high sugar diet will drive up your weight and make you obese[3]—just recall the familiar phrase, "You are what you eat." But this is only a fraction of what controls your weight and your metabolism. Indeed, conventional medical thinking completely ignores a large part of our new understanding of what controls the body's metabolism: the fact that your body has a rhythm, a circadian rhythm or a day night cycle[4]. It has an ebb and flow to its processes. These processes are not steady like the surface of a flowing stream, but are rather more like that of a cresting ocean wave—rising and falling throughout the day.

So how was this underlying metabolic pathway activated differently in these two mice? The answer was through their eating and fasting times. The researchers modified just this one very surprising environmental condition: *time*. That is,

---

[3] Barbara J Rolls, Phd, "Dietary management of obesity Cornerstones of healthy eating patterns "*Med Clin North Am.* 2018 Jan; 102(1): 107–124.

[4] Joseph S. Takahashi, "Transcriptional architecture of the mammalian circadian clock", *Nat Rev Genet.* 2017 Mar; 18(3): 164–179.

they changed the feeding times of the mice, which in turn built in a daily fasting window for one and not the other. They did so by leaving one mouse with a metaphorical open pantry within its cage, giving it free access to food around the clock, while the other mouse was given a pantry that only opened for a few hours a day, thus only giving it access to food for only a certain amount of time.

"By limiting the eating interval for one mouse and allowing an open eating interval for the other mouse, they were able to manipulate metabolism," I described to my conference audience. The lean mouse only had an 8 hour interval during which it could eat, and it therefore had a regular built-in nightly fasting window. Its feeding regime was based on the principles of Time Restricted Eating (TRE). Meanwhile, the feeding interval of the other mouse was not restricted in any way, and it was allowed to nibble throughout the day and night. This was the only environmental condition that was changed—a prolonged period of eating with almost no fasting window at all, versus a shortened interval of eating with a longer time spent fasting. It was this first restricted eating and fasting routine in the lean mouse which enabled a recalibration of its own internal timing system.

Researchers in the field of circadian biology used the mouse study I described to manipulate this silent timing system present within all animals. For these twin mice, the intervention was to "break" the internal timing system of one, while augmenting it in the other. Since both mice ate the same amounts of the same kinds of food, the one and only change in the environment between these two animals was *timing*. To reiterate: the only thing the researchers did to achieve this extreme outcome was to change the timing of *when* each of the twins were fed.

That mouse study was published by Dr. Satchin Panda's lab in 2012, and it broke conventional thinking in the medical community by defying the 11,000 previously published papers listing diet as the main influence on obesity. For those who were paying attention, the theory that diet and overconsumption were the sole causes of obesity was now and forever shattered. The study instead demonstrated that metabolism could be transformed by controlling the timing of eating to yield vastly different outcomes. This new discovery revealed that the time when a person ate was not only *a* critical factor that controlled metabolism, but it was *the* critical factor. The timing of eating could accentuate a circadian rhythm or dampen it, resulting in accelerating or suppressing metabolism.

Changing the time of when the lean mouse could eat revealed that a bypass route for metabolism existed, and that a back door or hidden mechanism resided within the body, this silent circadian rhythm. By ensuring well-organized eating and fasting intervals, the outcomes of any diet, regardless of how indulgent, can shift, and this opens up immense possibilities for your health.

Time Restricted Eating[5] can thus be defined further as the specific window of time when meals and snacks are eaten,

---

[5]   Amandine Chaix and Satchidananda Panda, "Time-restricted feeding prevents obesity and metabolic syndrome in mice lacking a circadian clock" *Cell Metab* 2019 Feb 5; 29(2): 303–319.e4.

as well as the specific window of time when fasting occurs, within a 24 hour day. This method of eating and fasting activates changes in the body's metabolism using this bypass route—one based on your internal circadian rhythm.

When first reported, the results of this animal study hit the media like the uncapping of a safety flare. Suddenly, people were attempting to follow the eating schedule from the mouse model. Terms like "16:8," "OMAD" (one meal a day), and "intermittent fasting" abounded in health and fitness circles. However, humans are not mice, and we have many other underlying mechanisms and pathways that benefit our health and offset some of the choices we make in our diets. Still, by harnessing some aspects of this and the many other studies from the circadian biology research that followed, the potential for humans exists on a similar scale to what the lean mouse achieved—all with eating within a TRE time frame[6], a TRE framework that is slightly more flexible and still works for humans.

This is an entirely brand new concept in medicine: that the intervals of eating and fasting can activate hidden metabolic pathways[7]. In fact, during my years under the

---

[6] Satchidananda Panda and Pam R. Taub, "Ten-hour time-restricted eating reduces weight, blood pressure, and atherogenic lipids in patients with metabolic syndrome" *Cell Metab.* 2020 Jan 7; 31(1): 92–104.e5.

[7] Satchidananda Panda, Nicolas Rodondi, Felix Naef, Tinh-Hai Collet, "The Effects of Time-Restricted Eating versus Standard Dietary Advice on Weight, Metabolic Health and the Consumption of Processed Food: A Pragmatic Randomised Controlled Trial in Community-Based Adults" *Nutrients.* 2021 Mar; 13(3): 1042.

collaborative guidance of Dr. Satchin Panda, we observed participants in a community clinical study on the use of TRE. What was discovered was that individualized eating and fasting intervals specific to a person's TRE window can sync circadian rhythms to the point of raising the chances of them achieving persistent, long-term metabolic improvements. Traditional "diets" based on calorie deficits result in 95% of participants gaining back their weight by year 5, but we already have patients breaking that statistic with TRE. In fact, because it is instead a way of resetting your internal rhythm, TRE is not actually a form of dieting at all. It just activates your own bypass metabolic pathway, which has always been there but has been silently suppressed. All of our successful participants in this ongoing metabolic study have outlasted even the best outcomes from any calorie restrictive or composition change diet.

The physicians in the audience had the same pressing questions in their minds as you do. They were wondering what exactly "Time Restricted Eating" is, and how it compared to Intermittent Fasting or the myriad of other fasting regimens that are offshoots of this original animal study. What are the differences between them? To explain all these different categories of eating and fasting regimens, I next shared a slide taken from the journal *Science* in 2016[8]. The article highlighted the many different "styles" of eating and fasting that the general public lump together as "intermittent fasting." All major eating and fasting regimens do have

---

[8]   Michel Bernier and Rafael de Cabo "A time to fast" *Science* 16 Nov 2018: Vol. 362, Issue 6416, pp. 770-775

their own specific benefits and attractions. However, they differ by duration and intensity, with each having its own characteristics.

In academic research, the metabolic changes caused by these regimens do not all fall under the lay term "intermittent fasting." Instead, they are actually broken down into three basic categories of eating and fasting cycles. Firstly, academic research uses intermittent fasting (IF) as a general catch-all term that refers to any prolonged period of fasting, typically greater than 24 hours. In contrast, TRE is defined by a daily interval of 8–12 hours eating and 12–16 hours of fasting. Another category, fasting mimicking diets (FMDs)[9] are defined by several days of an extremely low calorie diet followed by a period of fasting.

All of these categories have their own teams of researchers and scientists behind them, evaluating each model for cellular and biochemical level changes in animals. "TRE is a circadian or rhythmic form of eating and fasting," is how I explained it to the audience. "It has demonstrated the most persistent results from our own internal research data." Basically, when choosing a fasting and eating interval, TRE is the way to go for those who want disease improvement or even reversal, less medication and more weight loss.

I then steered my lecture into the topic of the health benefits of Time Restricted Eating: how TRE corrects and improves conditions like obesity, diabetes, hypertension,

---

[9]   Valter D. Longo and Satchidananda Panda, "Fasting, circadian rhythms, and time restricted feeding in healthy lifespan" *Cell Metab.* 2016 Jun 14; 23(6): 1048–1059

and fatty liver issues. "Once unleashed," as I expressed it, "TRE has the capacity to improve or correct all of these." I noted how "the results of our ongoing human study have shown the body revving up metabolism and turning on self-correcting mechanisms."[10] Our internal data continues to demonstrate unparalleled improvements in these chronic health conditions, in a way which was not only outlasting what a diet could do, but was even outperforming what medication could do. The most profound element of our data was—that we were not changing the *composition* of what our participants were eating at all.

The foundations of Time Restricted Eating arose from a field of research known as circadian biology—the study of the rhythms or timing systems built into all living things. You have most certainly observed these timing systems throughout nature: in a field of sunflowers that move their heads to face the sun throughout the day, to how a rooster typically crows in the early morning and not at night, or the squeaking of bats that only occurs at night. Nature built this timing system of circadian rhythm into all plants and animals, so that sunflowers know when to twist their blossoms, roosters are driven to crow at sunrise and not at sunset, and bats fly and forage at night rather than in the morning.

Your own internal timing system tracks the actual time of day, just as it does for plants and animals. This timing system is much more complex for humans of course, but you have

---

[10]   Julie Wei-Shatzel DO "The Benefits of Time Restricted Eating in Metabolic Symptoms" OPSC 30th Annual Fall Conference (Monterey, CA) Sept 22, 2019.

already experienced its force when you feel that invisible need for sleep typically at night, and the drive to work and play during the day. These are the rhythms to your day— times when you feel more alert and social as well as times when you feel more relaxed or sleepy. This internal sense of timing within you is driven by a circadian force or a timing system that is rolled up in the DNA of your cells.[11]

This group of hundreds of physicians in a San Diego conference room were among the first to learn about this new metabolic pathway, and to bring this information back to their patients. These discoveries relating to circadian rhythms are happening so quickly, and the human applications developing so rapidly, that they have not yet reached the medical textbooks. Even the most recent medical school graduates still may not know how to apply these new principles. That means you, the physicians who attended this conference, and those who have been attentive to the discoveries coming out of circadian rhythm research are among the earliest individuals to understand this new principle. If you embrace the lessons learned from this ongoing research, you will also be among the first to use them directly as a way of reaching a new level of health.

---

[11] Satchidananda Panda, John B. Hogenesch & Steve A. Kay "Circadian rhythms from flies to humans" *Nature* volume 417 329–335 (2002)

Chapter 2

The "Clock" In Your DNA

What are you made up of? What controls every aspect of how your body looks and functions? As you may already know, the answer is your DNA: a set of instructions (or "software") that is housed within every cell of your body. This DNA software is unique to you. It not only holds information related to your hair color, height, and ancestry, but it also holds the information needed to manage all of your bodily processes. For example, the production of all the proteins, enzymes, and transmitters needed for your brain and other organs to function are all controlled by the DNA within your cells. Your digestion, your nutrient uptake processes, and your ability to absorb vitamins and minerals are also determined by the way that the sets of instructions in your DNA are activated.

But there is something amazing about your DNA that you may not have known; something only discovered recently through this new field of research of circadian biology, the implications on human health are still only being revealed. The fact that, not only does DNA hold the genetic information that makes up all your different bits, but there is also a tiny bit of your DNA real estate that Mother Nature has reserved just for herself. Within this tiny space, she has inserted a sequence of instructions that has nothing to do with the usual instructions that determine your height and hair color, or the proteins that steer your internal processes. No, this insertion is nothing at all like the other sections of your DNA; it is something invisible and intangible, a code for something that keeps you forever linked to Mother Nature herself. This reserved segment of DNA tracks time at exactly 24 hour intervals, corresponding to one day.

Essentially, Mother Nature built a miniature time-keeping system right into every cell of your body—one that mimics the daily rising and setting of the sun. It is a timepiece that tracks the cycle of dawn and dusk; a template for a timer that tells your body when it is day and when it is night. This DNA real estate owned by Mother Nature guides the precise timing of the sculpting and dissolving of a particular "clock" like protein inside of your cells, and this process of "making and breaking" this "clock" protein tracks exactly with the 24-hour cycle of dusk and dawn.[12]

This sunrise and sunset template allows your body to differentiate between when it is daytime and when it is nighttime, and to adjust all of your internal processes accordingly. The activity of every cell in your body is organized by this sunrise to sunset schedule, which means

---

[12] Joseph S. Takahashi "Transcriptional architecture of the mammalian circadian clock" *Nature Reviews Genetics* volume 18 164-179 (2017)

all your tissues and organs—including your brain—are also organized to keep their every function separated into day and night categories. Knowing whether the sun is up or down is extremely important in coordinating all these processes. This circadian or daily rhythm is what will act as you secret bypass mechanism whenever you practice Time Restricted Eating

The foundation underlying TRE is that the times when you start and stop eating (or start fasting) should be tied to your own personal circadian rhythm. Setting this rhythm sets the pace for all the other processes taking place within your body. Your eating and fasting intervals over the course of a 24-hour day guide the timing of when major systems in your body turn on or off. To break it down even further, your body has basically two modes related to its daily circadian rhythm: a daytime mode and a nighttime mode. Like a toggle switch, the closer you can get your eating and fasting intervals to align with these two day and night modes, the more rhythmic your body becomes, and the more optimally it performs.

Your daily circadian rhythm relates to your day and your night modes: your daytime activities such as work, exercise, and even socializing typically happen during the day, while your rest, repair, and sleep phases typically happen at night. There is a reason why you, like most people, fall into this typical day-night pattern of activities: you are circadian. Your daytime rhythm drives you from within, and directs the preparation of your internal processes to perform daytime activities. Your nighttime rhythm likewise drives you from within, prompting your body to begin a different set of activities: those related to rest and repair.

These rhythms in your body are either augmented or suppressed, depending on your eating and fasting schedule. A TRE schedule brings you the closest you can get to syncing with your daily circadian cycle. How close you are to matching Mother Nature's set rhythm in your eating and fasting schedule then paces the rest of your body's processes. The further out you currently are from a rhythmic eating and fasting pattern, the more potential you have for disease and weight improvement. The closer you are, the more quickly you will see even further stepwise performance increases from your body.

Chapter 3

Reconnecting With Mother Nature's Rhythms

How do circadian rhythms and metabolism become disrupted in the first place? This has very little to do with what you are eating and everything to do with *when*—that is, the times you are eating. The pattern of when you eat and when you don't, come from a series of habits you have formed over the years. The first impressions that shaped your eating pattern began in your childhood, and they evolved into the current pattern you hold as an adult. By adulthood, your eating pattern has become set by the rhythm of your life. Your personal work and leisure times drive when you eat. What nature intends for you, however, is quite different from the way the modern world drives your schedule. If nature had it her way, then your eating rhythm would instead be driven by the rhythm of Mother Nature herself.

You can understand how Mother Nature intended to drive your rhythms of eating, fasting, activity, and rest by exploring communities where technology has had very little influence over eating and meal times. You can learn a lot just by looking at how the rhythms of nature drive the daily activities of these communities—most specifically, the times when they eat.

Of course, there are very few populations left on earth that still live in harmony with nature's rhythms and thus have their daily circadian rhythms augmented naturally. Among these few remaining communities are the Maasai tribes, living on the border between Tanzania and Zimbabwe. The rhythm of a Maasai's day begins and ends with perfect circadian timing, from the very first hour when they awaken and begin eating, to the time when they begin their nightly fast. Everything about their daily activity reinforces this regular rhythm of eating and fasting.

This is because they live very similarly to how their ancestors lived thousands of years ago. As herdsmen, they spend their entire day outside walking in the sunlight to find pasture for their cows. They are out at dawn and continue walking until dusk. The waning sun is their cue for going indoors, and their "indoors" are simple huts that they use only for sleep and rest. At all other times, they are soaking in the direct rays of the sun during the day. For hundreds of generations, they have also preserved strict adherence to a high protein diet, composed mainly of meat and milk. Consequently, adult Maasai males have an average BMI of 20[13], which is considered optimal. For

[13] Mbalilaki JA, Masesa Z, Strømme SB, Høstmark AT, Sundquist J, Wändell P, Rosengren A, Hellenius ML. Daily energy expenditure and cardiovascular risk in Masai, rural and urban Bantu Tanzanians. Br J Sports Med. 2010 Feb;44(2):121-6. doi: 10.1136/bjsm.2007.044966. Epub 2008 Jun 3

comparison, the latest statistics demonstrate that the average BMI of an American male is drawing close to 29!

For the Maasai, the times when they are eating, fasting, and performing other activities depend on the constraints that nature imposes. They have unknowingly mastered their eating and fasting intervals to match their circadian rhythm. They have fully unleashed their bypass metabolic pathways with their daily regular eating interval, which falls within TRE ranges every single day without fail.

The absence of artificial lighting, readily accessible food, and modern transportation, all create a daily rhythm for the Maasai that is very different from yours and mine. One obvious difference is that they are outdoors for as long as we are indoors during the day. They are only indoors after sunset—to sleep at night. This example of extreme synchronization with nature—with their eating, active, resting, and fasting times—provides us with a template for unlocking the full possibilities of what the human body can achieve.

The health and fitness of the Maasai have long inspired research teams[14] going as far back as the 1960s when researchers began to observe their diet and lifestyle. However, the synergy between their eating patterns and their circadian rhythms has only recently been recognized as likely playing a pivotal role in their robust health. Indeed, they are sometimes referred to as "the beautiful people" because they

---

[14] D.L. Christensen "Cardiorespiratory fitness and physical activity in Luo, Kamba, and Maasai of rural Kenya" *American Journal of Human Biology* 24:723–729 (2012).

are known for their clearly observable vitality, natural fitness, and lack of lifestyle related diseases.

For example, the Maasai are known to participate in feats of physical strength, such as in the "jumping dance." In a single motion, a young Maasai man launches his entire form up into the air with his upper body held perfectly upright, in a movement that makes him appear to fly upwards. He easily reaches up to 50 cm off of the ground—seemingly effortlessly. Could it indeed be that their rhythms of eating and fasting, of exercise and sleep, are major contributors to their high level of natural fitness?

Now that you know what the ultimate rhythmic lifestyle looks like in the Maasai, you can begin to see how the natural environment must play a major role in guiding any application of TRE. For the Maasai, nature's conditions dominate; their eating and rest periods are organized around the timing of the day and night. This pattern is also programmed into your genes. You may not be able to return to the rhythms of your ancestors, but by adjusting some of your major rhythms, you can still make dramatic improvements within your body's processes. TRE releases one of the foundational rhythms of the body from modern constraints: the daily rhythm of eating and fasting.

Within the walls of your home, from the comfort of your temperature controlled room, indoor and outdoor lighting, and with the convenience of a kitchen, it is difficult to begin to notice how many degrees of separation you are away from the original landscape and natural environment your body may actually be tuned to. This contrast between the pressures of modern life and the driving forces of nature is rather extreme for citizens of the modern world. Yet understanding how unusual these comforts and conveniences are is essential for you to see why TRE can have such a large positive impact on your health.

The decision to add TRE to your life is actually the first step in a commitment to aligning all of your body's rhythms, and is therefore more than just an eating and fasting regimen. TRE puts everything else into rhythm as well. Adhering to TRE aligns not just your digestive, nutrient uptake, and fasting intervals, but also creates the momentum of aligning all of the pathways within your body[15].

---

[15] Amandine Chaix, Emily N.C. Manoogian and Satchidananda Panda "Time-Restricted Eating to Prevent and Manage Chronic Metabolic Diseases" *Annu Rev Nutr.* 2019 Aug 21; 39: 291–315.

Chapter 4

Your New Daily Method

No matter where you begin on the spectrum of eating and fasting rhythms, you will see results from TRE[16]. Success is a matter of understanding how to properly implement TRE, how to recognize the markers of beneficial change, and finally how to tap into the method of persistence for ultimate health gains. By restoring your natural circadian rhythm, you will activate this important metabolic pathway. However, in order to fully utilize TRE, it first and foremost requires that you reflect on the current rhythm of your day. Evaluating how your personal circadian rhythm coincides with or diverges from the natural circadian environment around you is the initial step you must take.

TRE is a practice. It is not something that can be switched on in an instant and make a permanent change. It requires

---

[16]   Amandine Chaix, Amir Zarrinpar, Phuong Miu, and Satchidananda Panda "Time-restricted feeding is a preventative and therapeutic intervention against diverse nutritional challenges" *Cell Metab.* 2014 Dec 2; 20(6): 991–1005.

that you practice sticking to the eating and fasting windows you are committed to each day. Because you are really *committing* to setting a new rhythm of eating and fasting, you will want to keep it well-defined. This is a practice that takes only a few weeks for some to learn, while for others it can take a few months. Then there are those who push their TRE skills to the maximum—for this group, they continue to gain incremental health benefits even after years of practice.

I would like to share with you how to figure out the TRE interval that is ideal for your current lifestyle, and how you can continue to move up the ladder of TRE like those who have been practicing it for years. In this way, you can become a "Virtuoso" level TRE practitioner.

Your Personal Eating Style

To begin your own rhythm reset with TRE, the key is identifying where your current rhythm rests, which is based on your current eating habits. Through my years of working with patients applying TRE to their lives, I've discovered that almost everyone fits into a particular personal lifestyle of living and eating. Identifying your lifestyle and eating style will allow you to determine what new tools you will need in order to successfully adopt TRE into your life.

The Craftsman

The first lifestyle is that of the Craftsman. Just as she is skilled in her labor—effortlessly creative in all she does—the

Craftsman has the skills necessary to follow a set of somewhat complex instructions, as well as to navigate through her environment to ensure that she obtains her desired outcome. She can comfortably bypass and deflect away from whatever gets in her way, and is not deterred by distractions. She puts in the required effort at exactly the right time and the right place, allowing her to adhere to any new protocol and apply it to her life with ease. She is able to quickly adopt and adhere to TRE, and seems to enjoy the challenges that come with taking a new path. She sees this process as a way of adding a new skill to her toolbox. She is always ready to learn and, having already amassed a basic set of life skills, she fully adopts TRE for its clear and consistent benefits.

The challenge for the Craftsman when it comes to TRE is bridging its potential benefits with the amount of effort it takes to apply it. However, because the Craftsman is already quite adept at picking up new skills, the time it takes for her to "learn the practice of TRE" is a fraction of what it takes for others. She just needs a week or two of dedicated practice before she observes all the positive changes, and this reinforces her determination to proceed onward with this new regimen of eating times.

This is the ultimate goal of any eating style: reaching the level of the Craftsman. It is through single-minded determination that a Craftsman makes all the adjustments in their life necessary to adhere to a TRE schedule. Craftsmen are careful to not only finish dinner on time every evening, but they steer clear of any evening snacks, no matter how delicious or how strongly they are tempted.

The Craftsman - Jim & Karen

Jim and Karen both popped into view on my Zoom screen. They were sitting next to each other at home on their couch. This was our first interaction—I was covering the evening telemedicine shift, and they had requested an on-demand visit. Despite the limitations of such a telehealth visit, I could see the worry on Karen's face. She looked serious as she said, "He failed his physical today. If he doesn't pass the test, he won't be able to drive, and we'll lose our only income." She was referring to her husband's "DOT physical"—Jim is a professional truck driver. He drives 12 to 13 hours a day while on shift, and his day starts at 3:30 am. He must take the DOT exam every two years, and he had just failed it. He would have one more chance to pass at a repeat physical in less than two weeks, but this would be his last chance for a while. If he couldn't do it, he would lose his job and his paycheck.

"His blood pressure was 190 when they took it at his physical," Karen said. Her husband chimed in, "But when they retook it after a few minutes, I was down to 178." Clearly Jim was an optimist. But even with the few point improvement, his blood pressure had caused him to fail the fitness exam, he had missed passing by over 48 points. I looked through his medical records, and noticed that his last systolic blood pressure from his prior visit was highlighted in red on my screen: "180 systolic." This was close to the range he reported from his DOT physical. I then scrolled back through his visit history and saw that, at every visit to the office over the past two years, his blood pressure measurements had been flagged in red. Clearly, there had been a missed opportunity, but the challenge now was different from the challenge then.

Lowering blood pressure over time typically results in little to no side effects, but lowering blood pressure too rapidly—especially long-standing elevated blood pressure—can produce serious side effects. With his history of hypertension, the blood vessels in Jim's brain had adapted to a higher pressure, and alleviating this pressure too quickly could put him at risk of dizziness or—even worse—a syncopal event (a passing out spell). The brain is highly sensitive to its blood supply, and the layering on of medications too quickly could yield a side effect that would jeopardize Jim's license even further. He absolutely could not afford to have his blood pressure lowered rapidly with blood pressure medications—not while he was still actively driving for his job.

"You really gave me a challenge," I said. "We've got to lower Jim's blood pressure but we've only got 12 days to do it." Even though I was worried about slamming a blood pressure of 190 down to 130 in less than 12 days, I wasn't even sure that what they were asking for was possible. "Each blood pressure medication has the potential of lowering blood pressure by about 7 to 11 points," I said. "Jim, you need a 55 point drop to make it under the 140 marker." He looked serious, and I knew I didn't even have enough varieties of blood pressure medication to get him into that range.

"We've already been making a lot of changes for him," his wife thoughtfully responded. "In fact, he's cut down his fast food for the past month already, and he didn't have any fast food the week before the exam." She was letting me know that lifestyle changes hadn't seemed to have worked for him. It wasn't enough—his blood pressure clearly hadn't budged much from making those particular changes. But she also provided me with some critical information as well: Jim and she were Craftsmen, either each on their own or by their joint effort. As a team, they had already been working on a protocol to try to reduce Jim's blood pressure at home naturally. Karen was already working with him to make healthier diet choices, but they faced a typical roadblock—it was taking too long to have an effect on his blood pressure.

I sensed from this history they provided to me that Jim would be willing to make the right changes, and that if I provided them with the tools of TRE, they would implement them using their Craftsman skills. That was all I needed to know—it helped me immediately categorize Jim as having the right kind of can-do, will-do mindset, and to see that nothing was going to stop him. What's more, he had a wife who wanted to know what the real solution was, and who herself was ready to make any changes necessary. "I'm already furloughed from work, so I can make whatever needs to happen with our lifestyle happen," she told me. There was Craftsman commitment in her words. I still had my reservations since this was my first time meeting them, but I wondered, "Could Jim implement this in such a short time?" I certainly didn't want to guide him into something he wouldn't be able to participate in fully, but the consequences of hitting him hard with multiple medications within a week could be devastating.

I decided that I would coach him like a Craftsman—he had given me enough information already to see that he had all the traits. "We're not really going to change what you're eating, but we're going to do a little reorganization," I said. They looked at me a little puzzled—of course, neither of them knew what I was talking about—but I could see that they were committed to the task. I hoped that Jim would really bring out all of his Craftsman skills to adapt to TRE.

I briefly shared with them the new scientific understanding of metabolism and how it suggested an entirely different modality for blood pressure control. I shared Mark's story with the couple—a law enforcement officer who also had a blood pressure issue. His systolic blood pressure had been registering in the 160s and 170s, and he was on the verge of failing his fit-for-duty exam. Mark and I had worked together for a few weeks, moving his dessert up to a slightly earlier time and away from its initial time of 9 pm. By doing so, we were able to completely avoid the use of any medications to normalize his blood pressure. Mark not only passed his test back then, but he has maintained his blood pressure solely with TRE for over a decade.

Jim again revealed his Craftsman self. "Ok, tell me what I need to do and I'll do it," he said. He had a "no problem" attitude, and so he was ready to prepare for his test the right way. All he needed were the precise instructions. I brought his attention first to the "rules" of TRE, why it worked, and that a consistent time of finishing his dinner was of utmost importance.

We then began a detailed review of his whole day, from the start of his work day at 3:30 am to how he ended his day and his bedtime hour. I asked him to adjust his first cup of morning coffee—3:30 am was a bit too early based on his sleeping schedule—and then I asked him to move his dinner to a slightly earlier time, by about an hour to begin with. During that first visit, we had mapped out where the reorganization of his eating schedule would begin and where it would end. We developed a circadian pattern of eating for him that matched his swing shift work hours.

"We're going to reorganize your evening meal, move it up a bit," I told them. "Do you think you could do that?" Karen agreed, "I can make it happen, we eat dinner together early every night and I've got the time to make that happen." Again, her Craftsman side was telling me "no problem."

Jim was not a person who was obliged to eat due to external circumstances, and his wife was going to make sure of that. They were both committed to learning the steps and how to apply them. I never once heard from them any obstacles that they thought would keep them from fulfilling their plan to reorganize when they ate their meals and snacks. Karen also reassured me that she would make sure they fulfilled their commitment to stay on track, abiding down to the minute to the times we had discussed. Jim would have to give up on any snacks in the evening, and he would move his dinner up by an hour and a half. They considered it done on the first day we met, and there wasn't anything that was going to stop them from adhering to this new schedule.

Within the next few days, Karen had already begun initiating a meal prep plan, so that their dinners and lunches were already finalized the day before. Within 6 days of implementing this new, more circadian style of living, Jim texted me with his last three home blood pressure readings—they were now consistently lower at 136/81, 142/91, and 138/87. He had dropped his blood pressure by over 40 points just through TRE! This was the expected outcome for a Craftsman, and I knew he was now in the clear!

He trended in this range for the next few days, and I was assured that his blood pressure had now stabilized at this new normal. At this point, only the smallest amount of blood pressure medication was needed, as TRE has a synergizing effect with medication. I added the tiniest dose of one of the weakest and least side effect prone blood pressure medications (5 mg of amlodipine) and this yielded another 5 point reduction in his reading, easily slotting his blood pressure into the low 130s. He was now beautifully in range, and they texted me their delight when he passed his DOT exam with flying colors later that week.

The Compassionates

The next group—the Compassionates—are the kind of people we rely on in times of need, often acting as the metaphorical "rock" for a network of friends and family, but they too sometimes need to emotionally download. Since they are typically the constant emotional support for so many others in their community, food oftentimes becomes their own source of emotional comfort. Food, whether it be sweet or savory, is calming to the nervous system, and the Compassionates have discovered that it helps them to cope when their nervous system is in overdrive. The act of eating can help them to calm down, and alleviates both tension and irritability. This emotional eating relieves the deeper issue of an overly fatigued nervous system, which may struggle to adapt to a sometimes turbulent environment.

For the Compassionates, their journey on TRE is much longer, with some needing coaching at regular intervals for several months. They do best with consistent feedback and solid documentation of their incremental improvements. This is where the tracking app in our study really came into its own, as it provided our participants with game-changing data: daily, weekly, and monthly compilations of their efforts.

Compassionates often start at the wider end of a TRE schedule, closer to a 12 hour eating and 12 hour fasting interval, but they more and more begin to adapt to the traits of the Craftsman the longer they stay on track. It sometimes takes them longer to see the same level of changes but, as they wholeheartedly chip away at what needs to be done, they too will eventually begin to see big improvements. I am currently working with several Compassionates who have not only successfully kept their weight off for years, but who have already achieved some amount of fatty liver reversal. The greatest obstacle for these Compassionates is just to stick with it—their journey is longer, but the rewards are just as sweet.

For many Compassionates, they may not even realize that it is their own nervous system which is driving them to eat at any hour of the day. When trying TRE for the first time, a Compassionate feels like someone is snatching away their morning cup of coffee. They can really have a difficult time changing their eating patterns, regardless of the health benefits they know can be achieved.

The first move for a Compassionate is to identify that eating past the hour of 9 pm, regardless of the day of the week, is an indicator that food is being used as a tool to quell rising tension within their nervous system. Compassionates may even believe that TRE has no place in their lives—they would rather move on to a new diet than risk giving up their habit of eating late into the evening. They may even be successful at TRE for a week or several weeks, but their nervous system eventually rises up and overtakes their willpower, and they succumb to their drive to snack or imbibe late at night. They feel it as a pull from the inner core of their body, and changing this feels terrible. But if a Compassionate can learn to precisely adhere to TRE for those first few weeks, they will have overcome one of the largest hurdles in their life: their own resistance. This will make an indelible change within their own rhythm and, as they experience the changes within their body, they too will want TRE to remain a part of their life for the long run.

The Compassionate - Denise

Denise is a typical Compassionate. When she first came to see me, she was eating at nearly one to two hour intervals throughout the evening, and sometimes into the early morning. As a lead mental health counselor, her job involved giving all day, and she emotionally supported those who were in need of it the most. In fact, during the sars-cov-2 crisis, she continued to be physically present at the now-empty counseling center, and manned the phone line herself to make sure anyone calling in with a need would be supported.

She was not asked to do this, she just knew it was the right thing to do for her community. Even during our metabolic counseling visits, her cell phone rang constantly. It was always someone from her own personal community, calling her in need of a phone number, a resource, or just a few kind words. She was a constant supportive force in the lives of all of those with whom she counseled and worked with.

During our early sessions, the self tracking she was doing through the study app revealed food intake frequently occurring after midnight. In fact, a few nights a week, she would have eating intervals that ran from 4 am all the way to 2 am the next morning. She was essentially using food to keep herself from falling asleep, and as a distraction if she woke up. For Denise, getting to a regular 12 hour eating and fasting interval would not happen overnight—her body was already tuned to this erratic eating schedule. I knew that, as a Compassionate, her nervous system would likely retaliate if we shifted her eating times too quickly.

Instead, we started with something very simple: she would choose one time that she felt would not be too intimidating for her to stop eating, a time that she felt would be manageable to stick to every day until our next visit. So, Denise chose to start with midnight as her first manageable goal. This was not far from her current "free" style of eating, but it was nonetheless a certain (if small) parameter for her to aim at. It would take time for her nervous system to adapt, to not "talk back" to her about this new method, and indeed Denise was able to adjust gradually over a period of weeks to end her eating interval at midnight with 100% consistency.

The key for a Compassionate is to know that there is not one speed required for someone to adopt TRE successfully. The method can be adapted to any lifestyle, regardless of what the person's current circadian timing is like. Denise and I have been working together for a year, and she has now successfully moved the time of her last evening food item all the way back to 8 pm. She is now able to incorporate a small evening fasting interval into her day, and this has helped her in controlling her diabetes, her blood pressure, and her weight. She has managed to keep over 25 lbs of weight off without any of the yo-yo effect she had previously seen with the other diet regimens she had tried.

By adapting her entire lifestyle to TRE, and never again eating after midnight, Denise has seen significant improvements in her health. She is now consistently comfortable with ending her eating window at 8 pm, and she no longer needs the support of snacking through the night and into the early morning. She has noticed dramatic changes in her health, and the added perk of weight loss. And the beauty of TRE is that she still has room for improvement, depending on whether she now wants to move up to the next level; to that of a Virtuoso. This move is there waiting for her, whenever she is ready to take it on.

The Compassionate - Amanda

Another patient of mine had a similar story to tell. As a life coach, Amanda demonstrates all the characteristics of a Craftsman. She pays attention to details, her office is meticulously organized, and she follows a very regimented daily schedule for work. However, Amanda revealed to me that, even after eating her three meals a day, she might still feel hungry late into the night. She would often find herself nibbling on something close to midnight, and she had an erratic sleep schedule because she sometimes had trouble falling or staying asleep. In fact, her hunger sometimes woke her up in the early hours of the night, and she might be up snacking at 2 am or even 3 am if her severe insomnia had set in.

A bowl of cereal and milk was her go-to food. "I just love the creaminess of the milk with the crunch of the cereal," she told me. It wasn't that the cereal itself was a problem—there are no real problem foods in TRE—it was the time she was eating cereal that was the problem. She typically ate cereal close to her bedtime or when she awoke in the middle of the night. Rearranging the time when she had that bowl of cereal would be the single highest priority item for her to change her metabolism. By moving her favorite food to within a TRE window, I knew we would see the positive benefits in her health that Amanda was looking for.

Although she was a Craftsman in other aspects of her life, it was clear to me that Amanda was a Compassionate when it came to her eating schedule. She felt not only the burdens of her clients but also her own. It was both hunger and the comfort of food that drove her to snack late at night and early into the morning. "Something's got to fuel me," she said. "I didn't sleep for almost three nights in a row—I can tell you the exact dates." She was even organized when it came to tracking her personal data—she had jotted down the nights when she had barely slept, the days she ate late, and the off hours she kept. The "fuel" that helped Amanda run her otherwise organized day was food.

As a Compassionate, I could see that Amanda needed a different outlet for her stress. She carried a lot of silent stressors with her and, as with all Compassionates, these translated into her nervous system. The electrical wiring of the body—including the brain and all of the connections between it and the digestive tract—will trigger the release of the major stress hormone cortisol when under pressure. Because Amanda had put her nervous system under constant pressure, her body was putting out the extra stress hormone cortisol, driving up her appetite into the late hours of the night.

Once I identified that Amanda was indeed a Compassionate, we focused on methods she could use at home to quell her nervous system and reduce her overall cortisol secretion so that her appetite would not be continually triggered. The physiological response to hunger fueled by stress can be severe: the muscles tighten, the hunger pangs deepen, and there may even be an adrenaline surge causing the body to nearly tremble with discomfort. A single bite to eat will immediately dampen these symptoms, sometimes eliminating them within seconds.

We had to find a different tool that Amanda could use when her body was behaving this way—a way to avoid her silent stress building up to this level. Amanda used her Craftsman skills to come up with a list of possible solutions that she could use throughout the day so that she would not reach this point of hunger.

Here is the list she developed for herself:

Fuel myself with some food throughout the work day so that I don't have to eat a huge dinner;

Try to finish my work on the computer by 9 pm so that I can work on my other home chores at night instead;

Try to get to bed by at least 11 pm;

Write down my list of things to do tomorrow and things I feel like I've accomplished so I don't dwell on this when I'm trying to get to sleep;

Keep a hot water bottle at my bedside so that, if I wake up in the middle of the night, I can place it under my neck to relax my nervous system;

Try a warm shower before bedtime, and keep a hot bath ready that I can dunk my feet into to calm my nervous system in case I wake up in the middle of the night;

Try a gel mat for my bed—not heated or cooled, just room temperature to improve my sleeping comfort.

Amanda worked through her evenings, and slowly over the course of a number of weeks, she was able to adhere initially to a 9 pm limit for eating—with no snacks beyond that time. Once she accomplished this, she moved it up to 8 pm. She has found success now using her own Craftsman skills to shift her Compassionate side—by implementing her step-by-step solutions list on a daily and nightly basis.

The Entertainers

This type of person is a social butterfly, often entertaining or out with friends and family. Their gatherings usually begin in the evening and go on late, especially on weekends. They have a large network of friends, family, or both, and they often act as the connectors of a community. They enjoy being with people, and they often organize the gatherings. But Entertainers often feel the pressure to set the pace of imbibing and eating in order to make those around them feel comfortable. To be successful at TRE, Entertainers need to learn to shift into a Craftsman mindset in order to avoid the pitfalls of acting based on these social pressures.

The Entertainers - Wanda

My medical assistant Wanda has worked with me for nearly a decade. In the beginning, she saw the slow but steady transformation of patients who sought my help through my weight loss coaching or who desired to reverse their fatty liver disease. It wasn't until years later that Wanda decided to take

39

the plunge and implement TRE into her own lifestyle. She really didn't have weight to lose—in fact, she mainly made the shift because of the health benefits she was observing in the patients at our clinic. Wanda has a large network of friends and family, and many of her weekends are filled with family gatherings for birthdays, anniversaries, and graduations—a classic Entertainer lifestyle.

She shared with me her Craftsman solution to keeping up appearances and continuing to socialize through such evenings: "I just dump whatever is left in my wine glass at my [end of eating] time of 7 pm and fill it with water, then continue to enjoy my glass of water—which everyone assumes is still wine—and I go about my evening." Over time, Wanda has had such success with her weight loss that even our patients now comment on how healthy she looks. She has not changed her normal exercise routine consisting of walking her dogs, yet now patients coming to our clinic say to her, "Wanda, you are looking so good!" The main improvement they may not realize has been to Wanda's liver function, with her weight loss really being a side effect of that. She still manages both to hold her place as an Entertainer and to maintain her slimmer and healthier self by using her Craftsman techniques.

The challenge for the Entertainer of course is to maintain their regular eating windows despite the longer eating intervals of those around them. They must learn to find the Craftsman within them in order to navigate the difficult social pressures of eating and imbibing whenever they attend events. It does take some finessing, but it is even possible for them to improve their own health while also exerting a positive influence on those around them.

The Entertainer - Vivien

Another success story—Vivien—is someone who has learned to do just that. As a biodynamic farmer, Vivien and her husband regularly enjoy fresh garden fruits and vegetables. She is known by her friends for her homemade recipes, and as a generous farmer as well: she shares all that she grows with her friends and family. Vivien is an Entertainer—she has a large group of friends that gather regularly. For years, their dinners were always held at 7 pm. How Vivien was able to shift her lifestyle and her mindset from that of an Entertainer to that of a Craftsman made all the difference in her ability to maintain her goal weight over the long term.

Vivien appeared in a 2017 Wall Street Journal column describing how TRE is implemented, where she was featured because of the effortless way she was able to incorporate TRE into her busy social schedule. The article highlighted how Vivien no longer had to forgo the occasional dessert or other indulgence that she had previously so rigidly avoided. By adhering to TRE, those few extra nuisance pounds no longer accumulated. She used a dinner time modification— moving it forward by one hour—which meant she could enjoy the occasional indulgence without any consequences to her waistline.

Importantly, Vivien found success with TRE despite being an Entertainer. In fact, she found the method so compelling that she informed her monthly dinner club that she was moving their gatherings to an earlier time, and they all eventually followed suit. Being an Entertainer did not stop Vivien from bringing along those most important to her, and helping them to benefit from TRE as well. She and her friends are now living a social TRE lifestyle when they regularly gather on weekends. This has allowed Vivien to continue to enjoy her occasional indulgences along with her healthy cooking.

The Virtuosos

The next type—the Virtuoso—is the kind of person who has already implemented intense lifestyle changes over the years by eating right and exercising, and who is open to making any necessary changes to achieve continued improvements in their health. They believe in themselves, confident that they always make the most appropriate and beneficial health changes, and do not feel trapped by the sayings and doings of other people. They stick to their goals, plan out a timeline, and have a real target for when they expect to see results. Their results supersede any environmental pressures and they adjust their obligations to other people to accommodate their own parameters. The Virtuoso just needs the information and they begin applying TRE with consistency, as if they had been applying it for their entire life. In fact, a Virtuoso is normally already adhering to some version of a TRE lifestyle, and requires only a small adjustment to see even more gains.

I typically see a Virtuoso for only one to three visits. They are so accustomed to pivoting their lifestyle in order to achieve health improvements that they ease into TRE with little effort. For the Virtuoso, it is only a matter of time before they see large gains from adopting TRE. We generally don't need to track weight with Virtuosos, as they are usually only a few pounds away from (if not already at) their ideal. No—we track their liver health, and they are often able to achieve the ultimate heights here with a fully healthy liver.

Using blood test and ultrasound results, we can begin to observe how amazingly healthy the liver can become. For an organ that typically stores up years of metabolic burdens— no matter how healthy a person has been—it is wonderful to see how the liver health of a Virtuoso on TRE can get to the stage of liver age *reversal*. As liver activity changes at the cellular level, pathways for improving liver health occur through changes within the mitochondria, with cholesterol metabolism, and with blood sugar optimization. What's more, even brain health begins to shift in Virtuosos, as they begin producing increased levels of memory hormones called BDNFs. (brain derived natriuretic factor)[17]

Janet was a typical Virtuoso. She came into my office as one of our earliest metabolic study volunteers. As a high school PE teacher, fitness was her job, and she was always wearing tennis shoes and tennis shorts. She did however have one complaint: a nagging 10 lbs that she just couldn't keep off. Exercise kept her fit and her stamina high, but no matter

---

[17] Amandine Chaix, Emily N.C. Manoogian, Girish C. Melkani, and Satchidananda Panda, "Time-Restricted Eating to Prevent and Manage Chronic Metabolic Diseases", Annu Rev Nutr. 2019 Aug 21; 39: 291–315.

how many weekdays and weekends she spent playing tennis, it wasn't doing a thing for her weight loss goals. She found that she could forever be on a diet and live miserably to keep the 10 lbs off, or live normally, eat normally, and carry that bit of excess weight around. It was because of this nuisance weight, that as soon as she found out there was some new research on the horizon redefining how human metabolism works, she became highly interested in participating in our study.

When we evaluated her baseline eating pattern, the tracking app we used identified that Janet had her own unique rhythm of eating. First, she started eating at about the same time every morning, whether on weekdays or on weekends. With regularity she awoke at 6:30 am, and by 7:30am had her first cup of coffee. We typically see this kind of steady regularity in people's circadian rhythms: the time of coffee or breakfast is highly regular, with this regularity being maintained for some even on the weekends.

However, it was in Janet's evenings where the erratic patterns began. Although she finished dinner cleanly by 6:45 pm on weeknights, it was shifted an hour and a half later on the weekends, and she also regularly enjoyed a glass of milk later in the evening. When coaching our patients using the app, we measure intake start and end times down to the minute. The accuracy of timing really matters. Here it showed us that Janet was extending her eating window by that occasional glass of milk before bedtime, and by the irregular end times of her weekend dinners.

As a Virtuoso, Janet only needed to see her data. It was proof to her that her circadian rhythm was just a little bit out of sync on the weekends and on the nights when she had that additional glass of milk. This data and the idea of circadian synchronization was brand new information to Janet, showing her that it wasn't the intrinsic calories in what she was eating which was the problem but rather, these tiny variations in her eating habits that were disruptive to her circadian rhythm and thus to her metabolism.

Within three months of starting TRE, Janet had lost 12 lbs, and she had even kept it off with ease after an entire year. When I saw her later for a physical exam, she was delighted at her little "secret"—it had helped her keep to her ideal body weight and didn't require any further exercise or dietary changes. Janet the Virtuoso was reaping the full benefits of TRE.

The Obliged

This final type of person—the Obliged—cannot say no if offered something to eat or imbibe at gatherings. They are used to doing what everyone around them is doing. Even at home, they have difficulty prioritizing their time over others. Eating and mealtimes are not things directed by their own will or desires, but instead by what someone else—a spouse, child, partner, or other family member—requests or expects. They are in this way obliged to have their meals at times dictated to them by their environment, especially when it comes to dinner, snacks, and beverages.

The Obliged individual wants to achieve their weight loss or disease reversal goals, but does not even realize that it is their own inability to prioritize themselves which is the only real obstacle they face. The Obliged easily reveal their nature with statements like "I have to wait for my spouse to get home before I can eat dinner," or "We had a birthday party and cake was served late," or "It was my cousin's wedding and the food wasn't served until 8 pm."

The Obliged are extremely earnest in their efforts—in fact, they are typically a lot like a Virtuoso when it comes to how quickly they can adopt to a new eating regimen. However, if there is an event or an occasion to celebrate, or a friend suggests dinner at 8 pm, they are immediately thrown off track. The work of TRE requires consistency, so the Obliged has to direct all their effort into developing the support system they need to stay on track. Once the Obliged creates awareness in others regarding their reorganized eating regime, they find that they can follow TRE with ease.

Ruth was in just such a dilemma: she understood TRE, but it wasn't going to fit in with her husband's usual dinner time or with her family gatherings. She was in the mode of an Obliged, with her eating window set by her family and her husband. "It just won't work with what my family needs," is what she had told me years ago, but now we were having a different conversation. Now, we were discussing the drastic measure of adding insulin to her diabetes medication regimen.

I had already reached out to Ruth's Endocrinologist, asking if there was any other medication that she could take orally for her diabetes. The answer I got was just what I had expected: Ruth would need to start taking insulin. There simply wasn't any other drug that would give her the amount of blood sugar reduction that she needed. Her numbers were just too high; her blood sugar was running into the 190s in the morning before breakfast. Adding another oral medication would not counteract such a high blood sugar level.

We had reached a point where, after nearly a decade of diabetes, and after managing her blood sugar using only oral diabetes medications, Ruth's current treatment was no longer working. Ruth was set in her ways; she had lived with diabetes for over 10 years, but up until now had been unwilling to try anything as far as lifestyle changes go. She was facing what many of those suffering from type 2 diabetes face: the eventual loss of effectiveness of oral medications, and the need to add insulin injections to manage this next phase of their diabetes.

"I'll do anything but take that insulin!" she said. "You had mentioned there was something else I could do." She was referring to the suggestion of TRE we had talked about a few years prior. But back then she had clearly been in the Obliged mode, and had stayed in that mode for years. She obliged her husband's preference for late evening dinners. He was busy working in the garage and couldn't be bothered to come in early for dinner. She ate whenever her adult children and grandchildren arrived on weekends for potlucks. But now that her blood sugars were uncontrollable, and the medication had failed her, she was ready to switch modes. She changed her mindset from being the Obliged style of eating, to a Craftsman style. She asked the questions of a Craftsman. "I want the instructions on how to use the eating time changes," she said. She completely pivoted her role during this visit, and I could see that she was absolutely serious about using TRE to avoid "that insulin."

In all the years that I had known her, she hadn't been able to implement any new lifestyle changes at all. She was operating in the mode of the Obliged, and it seemed that she had been very content in that mode. But starting on that day, and for every day going forward, she completely changed directions. All she needed at that point were the exact instructions—she was after all a retired teacher. We evaluated her dinner time and chose 6 pm as a target for her to finish eating and to avoid any evening snacks after that.

Ruth didn't begin by counting calories or carbohydrates. Instead, she began by logging her food and beverage intake, and recording the time for everything she ate and drank. This allowed us to identify the nights when she was eating late. On average, she was starting breakfast between 10 am and 11 am, but it was her evenings that were more concerning. Some nights, her eating window ended closer to about 9 pm—especially if it was a bridge night with her friends. Because the target was her morning fasting blood sugar, we attempted her first adjustment in the evening by having her move her dinner up as early as she could manage.

In a dramatic turn, she managed to transform herself into a Craftsman within those three weeks. Facing the prospect of taking insulin, she kept to the plan by still attending the evening bridge club but avoiding the food. As she began moving her dinner time earlier to 7:30 pm, we already began to see changes in her morning fasting blood sugars. Moving from Obliged to Craftsman modes hadn't been worth the trouble for Ruth until now. Ruth would let nothing stand in the way of her making the right changes to avoid insulin. While she continued to track her morning blood sugars, we switched focus to what other food items and beverages could be shifted. Within another two weeks, her numbers began to improve even further. She was moving her morning glucose numbers from the 190s down to the 160s and, as the weeks progressed, she gradually saw her numbers return to the more normal ranges, fasting blood sugars ranges she hadn't seen in years.

Then Ruth came back to me with an entirely new end time that she had implemented on her own: she was closing her eating window at 5:30 pm. This was exceptionally early compared to her original baseline, but this was clearly her Craftsman mindset driving her to administer TRE for maximal benefit. Her logging efforts indeed demonstrated a clean stop at 5:30 pm without fail for two months, and her blood sugars had now returned to normal—without us adding a single additional medication. Not only was she completely without the need of insulin at this point, but we actually had to reduce her diabetes medications to compensate for the achievements in normalizing her blood sugar even further.

Ruth has been on this eating regimen for over five years now. "It's easy," she said to me during a recent visit. Even her adult children are amazed at how she sticks to finishing eating at the same time every night. At family gatherings, they ask her how she does it and Ruth tells them, "All I do is finish dinner by 5:30 pm, and that's all I have to do for my diabetes." Ruth essentially "saved herself" from having to use insulin; simply by starting TRE, stabilizing her blood sugar without changing a single thing about what she was eating.

Now her family, after seeing all the improvements she's made, work together to have their meals earlier whenever they gather. Ruth also hosts her bridge gatherings early to meet her eating time frame. She's still going strong on TRE after five years, happily attending bridge night without the disruption of having to administer insulin to manage her diabetes. "I'll never let this go," she says, referring to her permanent adoption of TRE into her lifestyle.

Chapter 5

Putting Your Rhythm In Sync

During my residency training years, my roommate was an Ob/Gyn resident. Her patients were pregnant mothers who needed care at all hours of the night. The deliveries and cesarean sections she was involved in were never limited to the day, and occurred just as often at night. She spent many working nights at the hospital, and would return home some days just as I was leaving to work my own hospital shift. Like most residents in training, her circadian rhythm was completely off schedule, but her medical specialty put her even more out of sync than many others.

She slept and ate her meals around her work schedule, which meant very little sleeping, and eating during most of the day and night. When she was home, she would eat at all hours. As long as she was awake, it didn't matter; she was always eating or nibbling something. As her roommate, I adopted this erratic eating habit along with her. For example, she ate chocolate at all hours of the day and night, and even kept an open bag of chocolate chips in the handy shelf door of the refrigerator. Several times a day (and often late into the evening), anytime one of us walked by the refrigerator, we would open the door and grab a handful of those chocolate chips. The time didn't matter. In fact, close to bedtime was exactly when most of the chocolate eating occurred!

This was a habit I kept for decades. As a young mother myself, each night after tucking the kids into bed, I would walk to the refrigerator, open the door, and scoop up a handful (or handfuls!) of chocolate chips. It was a reward, and it made me feel comfortable to have something special after a day of patient care, hospital calls, and running the children back and forth from daycare. Food, particularly sweets, helped me to relax in the evening, and chocolate was my go-to method for coping with my stress.

At the time, I had no idea I was living with the eating habits of a Compassionate, and it wasn't until years later that I began to understand the consequences of this bedtime snacking. My body had adapted to my years of erratic eating and was now trained to expect food at night—particularly sweets. I had provided my organs with the wrong timing template, with even my digestive system now adjusted to this late hour snacking. It took me months to learn the Craftsman skills of managing a new eating schedule and strictly adhering to it.

I still remember the first few nights when I began this shift in eating and snacking by completely stopping eating near bedtime. I felt extremely uncomfortable hunger pangs well up inside of me that made it feel as if I hadn't eaten for days! But because I expected this, I knew that it would be a short-lived experience—no more than a couple of weeks, or perhaps a little longer than that. Dr. Panda's research had already highlighted the expected symptoms of TRE, so I knew that the discomfort would eventually disappear as my body reset to a new rhythm. Those first nights were the most difficult because I was transforming my body's need for the reassurance of a food reward into a different need. I was teaching my body a new way to function in the evening; a new pattern of anticipating a short evening fast was now in motion. I was undoing years of the previous eating patterns and habits that my body had unnecessarily trained for.

By introducing this new regime, I was slowly teaching my body to expect a rhythm that would allow it to activate the repair and rejuvenation cycle it was originally wired for in the evening, and making it unlearn its expectation of food—specifically chocolate—late at night. I was breaking down my style of eating like a Compassionate, but I was not entirely destroying my Compassionate self. What I was doing instead was just driving this evening pathway down, and rearranging my chocolate cravings for a different time of day—for the daytime!

Ask anyone from my office and they will tell you where the dish of chocolates is located. It gets replenished multiple times per day by myself, as well as by a couple of other Compassionates in the office—there are just so many of us! The pace of work in a family medical clinic is a constant buzz: there are patients to serve, thousands of messages to return, appointments taking place in person and by video, and phone calls that require attention. To serve a community of patients well takes people like the Compassionates; their empathy and ability to connect with others allows them to thrive in situations of need. However, Compassionates need their small escapes, and our office chocolate dish is where you will see them coming through to take a few pieces for nibbling—including myself!

To successfully adopt TRE without completely swearing off the chocolate dish, I had to learn the skills of a Craftsman and find ways to stick to my set eating time frame. Doing so has allowed me to keep chocolate and the other foods I still enjoy indulging in, but they have just had to find a particular place within my eating window. For me specifically, that means eating no later than 5:45 pm. If I am at the office late, and having a conversation next to that chocolate dish, I will look up at the clock on the wall first before I take a few pieces. If it's the right time: bingo! I get my indulgence! But if it's the wrong time, well, I pass on it, knowing I can have them the next day.

The difficulty for any Compassionate is to adhere to this kind of rule, but consistency drives persistence. Importantly, because TRE does not involve any calorie deficits, I am able to continue to enjoy the foods I normally enjoy eating, it is just a matter of reorganizing the timing. Instead of creating a calorie deficit, I was creating a new rhythm of eating. TRE is able to accelerate metabolism because it resets your old rhythm to this new one; the one guided by Mother Nature. This metabolic reset only begins after weeks of consistency.

However, once I was able to achieve this reset, I knew my metabolism would begin to increase its pace as well.

Your pattern of eating is unique to you alone. You may already be close to your ideal TRE schedule, or you may still be a long shot from it. If the latter is true, you actually possess the greatest potential for improvements to your metabolism. If you are nearly perfect with TRE already, then you are in a great place to maximize the power and performance of your body. There is a place for TRE in everyone's life, whether you are normal weight, overweight, or just have a few pounds that you'd like to keep in check. TRE functions as a pathway for cleansing and rejuvenation, with weight loss just being a side effect of all the improvements to your body's performance.

Chapter 6

What to Expect

When you first begin adjusting your eating and fasting intervals, know that your body is still operating based on its previous experience, of years and possibly decades of irregular eating patterns, suppressing your true circadian potential. How far you are from your ideal TRE interval will determine how many levels you will need to move through in order to reach your goal rhythm. Every 15 minute interval makes a difference, so begin by deciding how many intervals you want to shift during the first week you dive into TRE. Determining up front what kind of eating pattern you currently have, determines where your current circadian rhythm rests. Your personal eating style will determine how rapidly or gradually you may want to adapt to your new TRE schedule.

A Craftsman typically wants to make large adjustments all at once. For example, Mark decided on a two hour adjustment to his eating window as an initial goal, and he achieved this within 6 weeks. Jim initially decided on a one hour adjustment, then moved to a two hour adjustment within his first two months. Craftsmen like these two quickly adapt and find it easy to persist—indeed, both Jim and Mark were able to stay on course with their TRE times even after several months on the program.

However, Craftsmen do not provide much feedback, so I can only assume that most of them still go through the same stages of discomfort that my Compassionates, Obliged, and Entertainers tell me about. The Craftsmen do seem to thrive on challenges, and any hunger pains they experience do not seem to discourage them from their goals. Once they have decided on the leap they are going to take, they don't need to make any more adjustments.

When it comes to the Compassionates, their tendency is to eat more frequently and late into the evenings. A Compassionate should begin with smaller, incremental changes. They are more likely to experience the challenge of wanting to break their evening fasting because of feelings of hunger, especially the longer they are awake late into the night.

In working with a Compassionate who struggled with night eating, Mary often felt hunger within only a couple of hours of finishing her dinner, and that made managing her chosen stop time each night a struggle. She wanted to dive enthusiastically into the program but, after the first couple of weeks, she was only meeting her goal time during about half of the days in a week. We decided that, even if she had to choose a later hour as her absolute ending time, it would still be to her advantage. After all, since she couldn't persist with her current goal, she wasn't actually setting her cell's timing system to a new rhythm at all. She decided to ease up on her TRE window and begin with a new time that she could manage easily every day. This strategy led to the start of a new and persistent rhythm for her.

The struggle for an Entertainer is finding a way to work TRE around social events. Rachel had tried TRE once in the past, and had entirely reversed her hypertension after 6 weeks—eliminating her medication in the process. But, by the next month, she was in my office again asking to be put back on the medication.

"I have to drink—everyone expects me to—and I can't keep skipping the drinking when I'm at a party." She had been right on target with her self description. Her eating style was Entertainer through and through and she wasn't willing to adopt any of the Craftsman skills to move her into a different rhythm. "I'm a social butterfly," she told me. She couldn't give up using alcohol outside of a TRE window, it just wasn't the time for her health as a priority. I reluctantly handed her the prescription she requested for blood pressure medication and she filled it that day.

Then, three years later, she came back to see me at the clinic and said she was now ready for TRE to again reverse her high blood pressure. "I'm retired now," she told me. She wanted to go off the medication for good this time. The experience of not needing medication and the lack of any side effects during that time had stayed with her. That memory was enough to propel her into using TRE with success this time around. But this time she put in the planning. She prepared herself to eat ahead of time for some of the events that she was still committed to. She also silently made a promise to herself to adhere to her alcohol cut-off time.

The Obliged seem to do best when they have the support of a spouse or friend who adopts TRE with them. In fact, when I see a couple come in for a visit together, I know that this will be a winning team. Having the support of another person in the household makes it twice as easy for the Obliged, with the shared experiences of the two people filling the gaps of expectation.

Knowing what to expect is half of the battle. For an individual Obliged working alone, they will have to rearrange some aspects of their environment to avoid the situations when they feel pressured into eating or imbibing. Vivien had found a work-around by inviting her friends over earlier— explaining to them that this was for health reasons—and it turned out to be a change that eventually all of her friends adjusted to. It even got to the extent that they started holding their own dinner parties earlier as well. In another case, Becky made arrangements to meet for happy hour with her friends, rather than getting together for their typical later evening dinners.

Chapter 7

Setting Your Rhythm

If you're a typical Compassionate, you eat at a normal dinner hour, but you may graze a little throughout the evening. When starting TRE, you will typically feel hunger in the evening, since it was your old pattern to nibble after dinner. It will certainly take some work to push through those first few weeks. It will be uncomfortable—but not for long! That's the trade-off you have to deal with; about three weeks of discomfort for lifelong health. By putting the work in during that time, eventually the hunger pains should lessen enough for you to scarcely notice them, and they should lessen even further as the weeks progress. How much discomfort you will feel will depend on the pattern of eating you had set for yourself previously. Rest assured that, as you move yourself into the correct eating and fasting pattern, you will create a new rhythm of hunger and satiety as your hormone levels move into this new pattern as well.

Your induction phase consists of weeks one and two, where you will want your palate to be completely clean of anything but water after your committed end of eating time in the evening. Keep in mind that the turning on of digestive juices and enzymes occurs with just one bite! Just one morsel of food does it, so you will want to completely avoid any snacking, regardless of how healthy or low calorie that food item may be. Remember that you are doing a full rhythm reset in order to set your new one. This means it is not the quality or quantity of the food that you have to score against; it is the time of when you finish.

How do you set your stop time? Begin by making it realistic. What we discovered in our weight loss study is that setting a realistic stop time is what turns TRE into a persistent habit. You do not need to go to any extreme. Remember that TRE starts with a 12-and-12 interval, and so if you are not consistently following this rhythm right now and for every single day, you will want to slowly move to it over the course of a few weeks. The latest goal time to begin your nightly fasting should normally be about 7 pm, but Compassionates may (due to their nature) find starting with a long fast in the evening a challenge. Still, you should remember that any change you make from your current baseline will be of benefit to your health. Even the smallest increment of 15 minutes can make a great deal of improvement—but your main goal has to be consistency!

If you have nights when you are eating or nibbling until 11 pm, and you make it your commitment to stop at 10 pm, then you should never go past 10 pm—ever! After a couple of weeks, dial it back to 9 pm, and stick to *that* for at least another two weeks. If you keep this stepwise pattern going, you'll be doing great. What you are doing here is first creating new patterns and developing new skills—particularly those of the Craftsman—so don't feel discouraged if you take more time than others.

Don't forget that your body—particularly your digestive system—has been allowed to run the show for you for years! Because of that, it has been taking advantage of the time that should have been allocated to other processes in your body—like housekeeping and metabolism. Even a one hour increase in your fasting time gives your body a precious chance to take care of these processes. This is how TRE helps you burn fat: in the final critical two hours before breaking your fast, your liver cells swell with fat, making it available for your body to burn.[18] Remembering this will help you adhere to your end of eating time and, if you must go past your committed end time, perhaps this will help you to stick as close to it as possible.

Once you have gotten through week two, continue with that momentum. These early weeks are for setting a rhythm; they are not the time to scrutinize your weight. I have had participants who dropped out within the first few weeks, right after they had set their new rhythm, because they did not yet see weight loss. They didn't understand that only once that rhythm is set will their body actively start doing its housekeeping work. This means weight loss is actually one of the last stages of TRE.

---

[18] Amandine Chaix, Amir Zarrinpar, Phuong Miu, and Satchidananda Panda, "Time-Restricted Feeding is a Preventative and Therapeutic Intervention against Diverse Nutritional Challenges" Cell Metab Vol 20;6:991-1005

By the time it gets to week 6, you will have formed a new habit, and you will now experience the beginnings of what it feels like to have a body working in synchrony. You will have embedded "the new you" into your cells by setting a proper rhythm to your eating and fasting times. You will have also re-worked when your cells will prepare to release hormones, enzymes, and proteins, and this will have given your body a new sense of anticipation.

It is by approximately week 8 that weight loss begins—although this will vary depending on the condition of your liver. If it has already succumbed to quite a bit of fat storage, your body will first focus on melting that liver fat away before it goes for the stores in the other areas of the body you might desire. Your body wants to do what's best for you, and your organs are the priority. Needless fat stored on the organs impairs their function, so that the organs will be targeted for repair and fat removal before any peripherally stored fat.

This benefits you in several ways: your metabolism rises, you feel more energetic, you sleep better, and everything works better. So don't be frustrated if you don't see accelerated weight loss in the early weeks. Your body is just finally getting its full breath back, and has started doing what it has always wanted to do for you—to fix everything from the inside out. TRE is a journey, don't feel rushed but persist on this new path and you will soon experience the transformation happening both on the inside and the outside. It is your natural rhythm, the rhythm that Mother Nature had always intended for you.

There is huge power in a change of just one hour.